THE AI THAT LEARNS FROM ITS MISTAKES
Google DeepMind's Revolutionary Leap

Exploring How Artificial Intelligence Is Becoming Smarter, Faster, and More Independent

Alejandro S. Diego

Copyright © Alejandro S. Diego, 2024.

All rights reserved. No part of this publication may be reproduced, distributed, or transmitted in any form or by any means, including photocopying, recording, or other electronic or mechanical methods, without the prior written permission of the publisher, except in the case of brief quotations embodied in critical reviews and certain other noncommercial uses permitted by copyright law.

Table of Contents

Introduction 3
Chapter 1: The Problem AI Faces: Inability to Self-Correct 8
Chapter 2: Enter Google DeepMind: The Rise of Self-Correcting AI 16
Chapter 3: How SCoRe Works: Breaking Down the Method 26
Chapter 4: Real-World Results: AI Gets Smarter, Faster, and More Independent 37
Chapter 5: Generalizing Self-Correction Across Multiple Domains 49
Chapter 6: Overcoming the Challenges 64
Chapter 7: The Broader Implications of Self-Correcting AI 77
Chapter 8: A World with Autonomous AI: What's Next? 90
Conclusion 105

Introduction

Artificial intelligence has developed at a remarkable pace over recent decades, transforming industries and redefining the boundaries of what machines can achieve. But amid these advancements, something extraordinary is happening at Google DeepMind that promises to change the way we perceive intelligence itself. Up until now, AI has been celebrated for its ability to perform complex tasks, often exceeding human capabilities in speed and precision. Yet, despite its power, AI has struggled with a fundamental limitation—it lacks the ability to recognize and correct its own mistakes. For the first time, however, that boundary is being pushed. Google DeepMind has introduced a breakthrough that enables machines to not only carry out tasks but also to learn from their errors and improve autonomously. This innovation represents a significant leap forward in AI's potential.

To truly understand the magnitude of this development, it's important to look at how artificial intelligence has evolved over time. AI's journey has been filled with remarkable achievements, from mastering chess to creating art, and from optimizing logistical systems to providing customer support. Yet, these successes have been largely confined to tasks where clear instructions, rules, and goals are predefined. The intelligence displayed by these systems, while impressive, has always been reactive, not proactive. When AI makes mistakes—whether in solving a math problem, generating code, or interpreting complex data—it relies on human intervention to identify and rectify the error. This reliance creates a bottleneck in AI's effectiveness, particularly in situations requiring multi-step reasoning. Without the ability to reflect on its own output and revise it accordingly, AI remains limited, even when it possesses the knowledge to get things right.

Enter Google DeepMind's groundbreaking method: Self-Correction via Reinforcement Learning, or SCoRe. This new approach has opened a door to an entirely new level of machine intelligence. Through SCoRe, AI can now independently identify when it has made an error and take steps to correct it without the need for human guidance. This method not only enhances the capabilities of AI but also pushes the technology closer to true autonomy, where machines can function, learn, and evolve in real time, free from constant human oversight. What sets SCoRe apart is its reliance on reinforcement learning. Unlike traditional AI models that depend on massive datasets and human-led supervision to improve, SCoRe allows AI to generate its own data, learn from it, and improve iteratively. This marks a shift in how machines learn, transforming them from passive recipients of knowledge into active participants in their development.

This book delves into this revolutionary leap in AI technology, exploring the methods behind Google DeepMind's SCoRe, its impact on artificial intelligence, and the broader implications for the future. Through this analysis, we will unpack how SCoRe is positioning AI to become not only more intelligent but also faster, more adaptable, and capable of taking on increasingly complex tasks without human intervention. The ability for machines to learn from their mistakes is a crucial step toward creating systems that can be trusted to operate autonomously in environments where precision and efficiency are paramount. From industries like software development to sectors like healthcare and education, the potential uses for self-correcting AI are vast and transformative.

By the end of this exploration, readers will have a comprehensive understanding of how Google DeepMind's breakthrough is paving the way for a new era in AI development, one where machines aren't just tools that need constant refining but partners capable of learning, adapting, and evolving in real time.

Chapter 1: The Problem AI Faces: Inability to Self-Correct

Despite the immense advancements in artificial intelligence, traditional AI models continue to grapple with a fundamental flaw: they struggle with multi-step reasoning. These models, even those built on vast datasets and intricate algorithms, often falter when tasked with problems that require multiple steps to solve. This isn't because the AI lacks the knowledge; rather, it's due to its inability to dynamically apply that knowledge across a sequence of tasks. A mistake made early on in a process can lead to a cascade of errors, with the AI unable to recognize and correct these mistakes on its own. This often results in an entirely incorrect final result, despite the AI having all the necessary components to get it right.

Imagine solving a complex math problem. You might be confident in the steps, but if you make a small error in one calculation early on, the entire solution can unravel. However, as a human, you have the ability to spot that mistake, backtrack, and correct it before continuing. You understand the larger context of the problem and can adjust accordingly. AI, in contrast, lacks that same awareness. Traditional models will continue executing the task, oblivious to any error, and will ultimately produce a wrong solution. This flaw is particularly problematic in tasks requiring layers of reasoning, such as coding or scientific research, where one wrong turn can derail the entire outcome.

To further illustrate this, think about debugging a piece of code. A small typo, like an extra space or a misplaced symbol, can cause the program to fail. As a human coder, you instinctively go back, line by line, to find and fix the mistake. You can recognize where the logic broke down and how to rectify it. AI, on the other hand, follows its instructions precisely but without the ability to recognize when something has gone wrong. It executes the code exactly as written, but if there's an error, it will persist unless someone intervenes. Traditional AI models are programmed to do exactly what they are told, even when the instructions lead to a mistake.

This inability to self-correct becomes even more evident in complex, multi-step tasks. A great example is in large-scale data analysis, where an AI might be tasked with processing millions of data points. If it misinterprets a key metric or misclassifies data early in the process, the final output will reflect that error throughout. Without the capacity to recognize and address that initial mistake, the AI continues blindly, compounding the error across all subsequent steps. It can become stuck in a loop of ineffective corrections or minimal changes that don't truly solve the underlying issue.

These challenges highlight the inherent limitation of traditional AI: even with vast knowledge and computational power, it lacks the critical ability to recognize its own mistakes and adjust its approach. This is where Google DeepMind's revolutionary SCoRe method becomes so vital, introducing a system where AI can now learn from its mistakes, course-correct, and ensure that each step in the process is accurate before moving forward.

Current approaches to addressing the flaws in AI reasoning have primarily relied on methods like prompt-based adjustments and supervised learning. While these techniques have seen success in controlled environments, they often fall short when applied to dynamic, real-world applications that demand more flexibility and adaptability from AI systems.

Prompt-based adjustments, for example, involve giving the AI model specific instructions to guide it toward the correct answer. This method works well when the task at hand is relatively straightforward or when the AI has enough information to follow a simple set of steps. However, when tasks require deeper reasoning or a series of complex decisions, prompt-based adjustments become less effective. The AI can still get stuck, unable to deviate from its predefined path even when it encounters an error.

The lack of adaptability in real-time means that AI is often unable to address mistakes that arise from unexpected inputs or novel scenarios.

Another widely used method is supervised learning, where AI models are trained using large datasets labeled with the correct answers. In this approach, the AI learns by example, memorizing patterns and applying them to future tasks. While supervised learning has been instrumental in advancing AI's capabilities, it suffers from a significant drawback—it's rigidly tied to the data it was trained on. If the AI encounters a situation that differs even slightly from the scenarios present in the training data, it can struggle to make accurate decisions. This method often amplifies biases or mistakes embedded within the original dataset, causing the AI to make shallow corrections or, in some cases, introduce entirely new errors.

Supervised learning also requires constant human intervention. AI models need to be regularly fine-tuned by developers to ensure they stay relevant and accurate as new data becomes available. This process is both time-consuming and resource-intensive, making it difficult to scale for real-world applications where conditions are constantly changing. Additionally, supervised learning doesn't provide AI with the tools to recognize or fix its own mistakes. Instead, the model relies on predefined correction paths or, worse, it may not correct the error at all, leaving the task incomplete or inaccurate.

In more complex tasks, such as programming or scientific reasoning, relying on prompt-based adjustments or supervised learning alone can prove problematic. These methods often result in models that either overcorrect by introducing unnecessary changes or underperform by failing to make meaningful adjustments. In programming tasks, for example, prompt-based systems might fix a small syntactic error without addressing a deeper logical issue in the code, leading to faulty output despite the AI's superficial "fix." Similarly, supervised learning may lead the AI to apply an incorrect solution if the task doesn't fit the strict patterns it was trained on.

In dynamic, real-world environments—whether it's financial modeling, software development, or even healthcare applications—AI must be capable of handling unforeseen challenges with minimal oversight. This requires a more flexible approach than prompt-based tweaks or supervised corrections can offer. Real-world data is messy and unpredictable, and AI systems need the ability to adapt on the fly, something that these traditional methods simply cannot provide in an effective way.

These limitations are precisely what Google DeepMind sought to overcome with its SCoRe method. By introducing self-correction through reinforcement learning, the team moved beyond the constraints of prompt-based and supervised methods. Instead of relying on pre-labeled data or manually inserted prompts, SCoRe allows the AI to generate its own data and learn from it, creating a model that can adapt and correct itself dynamically, even in unpredictable real-world situations. This breakthrough is a critical step in moving AI closer to true autonomy, where it can not only perform tasks but also recognize, learn from, and fix its mistakes without human intervention.

Chapter 2: Enter Google DeepMind: The Rise of Self-Correcting AI

Google DeepMind's SCoRe, or Self-Correction via Reinforcement Learning, represents a groundbreaking solution to one of AI's most persistent challenges: the inability to correct its own mistakes. Until now, AI models have relied heavily on external supervision or pre-programmed correction mechanisms, which, while effective in controlled environments, often fall short in real-world applications that demand flexibility and adaptability. SCoRe changes this paradigm by introducing a new way for AI to learn—through reinforcement learning. This novel approach allows AI to recognize its own errors and correct them dynamically, without needing human intervention or vast amounts of pre-labeled data.

What makes SCoRe revolutionary is that it bypasses the traditional methods of improving AI performance, which usually involve supervised learning. In a supervised learning model, the AI is trained on large datasets filled with examples, with each input meticulously labeled to help the AI learn the "right" answers. While this process can yield impressive results, it has inherent limitations, particularly in its rigidity. AI models trained this way are often constrained by the boundaries of their training data. When confronted with a situation or a problem that falls outside those boundaries, the model is likely to falter, unable to adjust to the new context. This is where SCoRe's reinforcement learning offers a significant leap forward.

Reinforcement learning works by allowing the AI to engage in trial and error. Instead of relying on pre-defined correct answers, the AI interacts with its environment, makes decisions, and receives feedback in the form of rewards or penalties based on its performance. Through this process, the AI gradually learns which actions lead to successful outcomes and which do not. In the context of SCoRe, this means that the AI isn't simply memorizing solutions—it's actively learning how to solve problems by correcting its own mistakes over time. The reinforcement learning framework enables the AI to generate its own data, use it to improve, and adapt its approach in real time, making it far more versatile than traditional models.

The true novelty of SCoRe lies in its ability to take this learning process a step further. Not only does the AI learn to identify errors in its outputs, but it also develops strategies to correct those errors in meaningful ways. This is a significant departure from older methods, where AI models might only make superficial adjustments, such as tweaking minor details that don't fundamentally alter the outcome. With SCoRe, the AI is designed to make deep, substantive corrections that address the root cause of the problem, ensuring that its self-improvements are both effective and efficient.

Moreover, SCoRe's reliance on self-generated data drastically reduces the need for external oversight, making the entire process more scalable. In traditional AI systems, human intervention is often required at multiple stages, whether it's in the form of correcting errors, fine-tuning the model, or providing labeled data for further training. SCoRe minimizes these needs by giving the AI the autonomy to correct itself, which not only makes it more efficient but also opens up possibilities for deploying AI in environments where human oversight is limited or impractical.

This ability to self-correct also addresses a key limitation in AI's application to complex, multi-step tasks. Previously, when AI encountered an error early in a task, it would often continue with the wrong assumptions, leading to a flawed final result. With SCoRe, the AI can detect these early errors and correct them before they propagate, dramatically improving its performance in tasks that require multiple layers of reasoning.

By embracing reinforcement learning and focusing on self-generated data, Google DeepMind has set the stage for a new era in AI development. SCoRe positions AI to become not just a tool that follows instructions but an active problem solver capable of learning from its own mistakes and improving over time. This breakthrough represents a major leap toward truly autonomous AI systems that can function reliably in dynamic, real-world environments, making them smarter, faster, and more independent than ever before.

The SCoRe method operates through a two-stage training process, each of which plays a crucial role in enabling AI to make meaningful corrections and improve its overall performance. This dual approach ensures that AI doesn't just tweak small, inconsequential details but instead learns to identify and fix the root causes of errors in a way that boosts accuracy and effectiveness.

In the first stage of training, the focus is on teaching the AI how to make significant corrections rather than just minor adjustments. Traditional AI models often fall into the trap of making shallow edits—changing a word here or tweaking a number there—without addressing the underlying problem. This can result in what seems like a corrected output, but in reality, the AI has only made superficial changes that don't improve the overall accuracy of the solution. SCoRe's first stage is designed to break this habit. The AI is trained to recognize when a deeper, more impactful correction is needed, allowing it to avoid getting stuck on surface-level fixes that don't solve the actual issue. This stage is essential for building a robust correction strategy, ensuring that the AI knows when to make small changes and when a more substantial revision is required.

To illustrate this, consider how a human might edit a flawed solution to a math problem. If you notice a calculation error, you don't just change the incorrect number—you revisit the entire reasoning process to ensure that every step is accurate. SCoRe teaches AI to follow this same logic, encouraging it to look beyond minor edits and assess whether a larger correction is necessary to fix the problem at its core. This first stage is all about laying the foundation for meaningful self-correction, equipping the AI with the ability to make thoughtful adjustments that go beyond cosmetic tweaks.

Once the AI has mastered the art of meaningful corrections, the second stage of training comes into play. This phase uses what is known as multi-turn reinforcement learning, a process where the AI is rewarded for improving its performance over successive attempts. In this stage, the AI is not just making one correction and moving on; it engages in a cycle of self-improvement, where each pass through the task provides an opportunity to learn from its previous mistakes and enhance its accuracy. The key to this stage is the reward system. By carefully shaping the rewards, Google DeepMind ensures that the AI is motivated to aim for higher accuracy with each iteration. Instead of receiving rewards for making superficial changes, the AI is only rewarded when its corrections lead to a substantial improvement in the final result.

This reward system is crucial because it helps the AI develop a more strategic approach to self-correction. In traditional models, AI might make a series of small, incremental changes that don't significantly improve the output. However, with the multi-turn reinforcement learning approach, the AI learns that it's not enough to make small adjustments. It must aim to make meaningful, effective corrections that lead to a better overall outcome. Each time the AI revisits the task, it evaluates its previous response, identifies where it can improve, and makes changes accordingly. This process continues through multiple iterations, with the AI receiving feedback after each pass, helping it refine its approach and become more accurate over time.

An example of this process can be seen in how SCoRe handles complex tasks like coding or mathematical reasoning. On its first attempt, the AI might produce a solution with a few minor errors. Instead of stopping there, the AI re-evaluates its work, identifying where it went wrong and making necessary corrections. After each pass, the AI becomes better at detecting the mistakes and revising them, with the reward system reinforcing its progress. This multi-turn process ensures that the AI not only corrects its mistakes but also learns from them, becoming more efficient and accurate with each attempt.

Together, these two stages form the backbone of the SCoRe method. The first stage ensures that the AI focuses on meaningful corrections, while the second stage, with its multi-turn reinforcement learning, allows the AI to refine its approach and improve its performance incrementally. This dual process enables AI to go beyond mere error detection and correction—it fosters a cycle of continuous improvement, where the AI learns to self-correct dynamically, evolving with each iteration. As a result, SCoRe produces AI systems that are not only capable of fixing their own mistakes but also of becoming smarter, faster, and more reliable with every pass through a task.

Chapter 3: How SCoRe Works: Breaking Down the Method

The core of Google DeepMind's SCoRe method lies in its innovative use of reinforcement learning to enable AI to correct its own mistakes. Reinforcement learning, at its essence, is a type of machine learning where an agent (in this case, the AI) interacts with its environment, takes actions, and receives feedback in the form of rewards or penalties. Through repeated interactions, the AI learns which actions lead to successful outcomes and which do not, refining its decision-making process over time. This contrasts with traditional supervised learning, where AI models are trained on large datasets filled with pre-labeled examples and then tested against those patterns.

SCoRe's brilliance comes from its ability to teach AI not just to perform tasks but to improve with every iteration by using its own outputs as data for further learning. Unlike supervised learning, where an AI model might depend on static datasets,

SCoRe enables AI to generate its own data in real time. This self-generated data forms the foundation for the AI's learning process. Each time the AI attempts to solve a problem or complete a task, it evaluates its performance, detects any errors, and then uses those errors as feedback to adjust its strategy. The AI is rewarded for making corrections that lead to a more accurate outcome and penalized when the changes are insignificant or incorrect.

This process of self-correction is particularly effective because it allows the AI to engage in trial and error without needing human intervention. Over multiple iterations, the AI becomes better at recognizing not only that it has made a mistake but also understanding how to fix it. In each iteration, the AI refines its strategy, learning which corrections lead to meaningful improvements and which do not. This cyclical process of generating data, learning from it, and adjusting based on feedback allows the AI to continually enhance its performance in a more flexible and dynamic way than what is possible with traditional methods.

The use of self-generated data is critical to SCoRe's ability to bypass a significant limitation in traditional AI models: the reliance on external data. In supervised learning, AI systems are only as good as the data they are trained on. If the training data contains biases or inaccuracies, the AI inevitably inherits these flaws, perpetuating them in its outputs. This is especially problematic in real-world scenarios where the data available for training often reflects historical patterns that may no longer be relevant or accurate. Furthermore, when the AI encounters a scenario that diverges from the data it was trained on, it often struggles to adapt, as it lacks the flexibility to handle unfamiliar inputs.

SCoRe addresses this problem by allowing the AI to work with self-generated data, which eliminates the dependence on external sources. This independence from static, potentially biased datasets significantly reduces the risk of the AI making shallow or biased corrections. Since the AI is constantly generating new data through its interactions and learning from its own mistakes, it develops a more nuanced understanding of the task at hand. The result is a more adaptable system that can apply its learning dynamically across different domains and situations, even those not covered by traditional training data.

Another key advantage of using reinforcement learning and self-generated data is that it helps the AI avoid the amplification of biases commonly found in supervised learning. When an AI model is trained on a biased dataset, it not only inherits those biases but often reinforces them, as the corrections it makes are based on flawed information. This can lead to AI systems that make decisions based on skewed or inaccurate patterns, causing issues in fields where impartiality is crucial, such as hiring algorithms, legal judgments, or financial modeling.

By allowing the AI to create its own learning environment, SCoRe circumvents these pitfalls. The model isn't locked into the biases of the data it was initially trained on; instead, it can adapt and evolve by learning from its own experiences. This ability to generate and correct its own data means that the AI is no longer beholden to the limitations of its training set. It can learn to identify and correct errors without perpetuating the biases inherent in external datasets. Moreover, as the AI refines its understanding of a problem through reinforcement learning, it becomes better equipped to handle real-world scenarios, where data is often messy, incomplete, or contradictory.

In practical terms, this ability to learn from self-generated data and overcome biases has far-reaching implications. AI models that use SCoRe are not only more accurate but also more robust in handling complex, dynamic tasks. Whether it's coding, mathematical problem-solving, or even tasks that require reasoning across multiple domains, AI systems built with SCoRe can operate more independently and more fairly, making decisions based on their own iterative improvements rather than on potentially flawed human inputs.

The combination of reinforcement learning, self-correction, and the generation of independent data makes SCoRe a powerful method for advancing AI. By breaking free from the limitations of traditional supervised learning, it allows AI to improve continuously, make fewer mistakes, and deliver more reliable, unbiased results. This flexibility and adaptability are key to the future of AI, pushing it toward true autonomy in real-world applications where the ability to learn from one's own mistakes is essential.

One of the most significant advantages of the SCoRe method is its ability to reduce computational costs while enhancing efficiency in AI's decision-making processes. Traditional AI models often require a second model or external system to verify and validate their outputs. This two-model approach is both resource-intensive and time-consuming, particularly in tasks that involve complex reasoning or multi-step processes. By contrast, SCoRe eliminates the need for this additional model by enabling the AI to correct itself through reinforcement learning, significantly reducing the computational burden.

In conventional AI systems, once a model produces an output—whether it's a solution to a problem or a response to a prompt—another model, sometimes referred to as a verifier, is employed to check the accuracy of that output. This adds another layer of processing, as the verifier must analyze the original model's work and determine whether corrections are needed. This approach, while effective in certain scenarios, is computationally expensive, especially when dealing with large-scale tasks or real-time applications. Maintaining two models to check and verify outputs increases the overall complexity of the system, requires more computational power, and slows down the entire process. Moreover, this method doesn't always guarantee that the corrections will address deeper issues since both models are often bound by the same limitations and biases.

SCoRe overcomes these inefficiencies by allowing the AI to function independently without requiring a second model for verification. Instead, the model learns from its own mistakes using reinforcement learning. It self-corrects by iterating on its outputs, using its own performance data to improve over time. This self-sufficiency means that the AI no longer has to rely on external validation, thereby streamlining the entire process. The result is a much more efficient system, where computational resources are focused on the task at hand rather than split between two models.

Another way SCoRe improves efficiency is by handling the mismatches that often arise between training data and real-world inputs. In traditional AI systems, the model is trained on a large, static dataset that provides examples of how to solve specific problems. However, these datasets, no matter how extensive, can never fully capture the diversity and unpredictability of real-world data. When an AI model encounters a situation that doesn't align perfectly with its training data, it struggles to adapt, leading to errors or incomplete solutions. The mismatch between what the model has been trained to recognize and what it actually encounters in practice is a common issue in AI, often requiring additional fine-tuning or retraining to address.

SCoRe offers a solution by allowing the model to learn dynamically from the data it generates during real-time interactions. Instead of being confined to the patterns and biases embedded in its initial training dataset, the AI can adjust its approach based on the actual data it encounters. Through reinforcement learning, the model receives feedback from its environment, evaluates its mistakes, and iteratively improves its responses.

This flexibility makes the AI far more adaptable to real-world situations, where inputs are often unexpected or deviate from the norm.

By continuously refining its understanding and correcting its own errors, SCoRe effectively reduces the need for costly retraining or manual fine-tuning. In scenarios where traditional models might need to be re-calibrated with new data to handle unforeseen inputs, SCoRe-trained models can adjust on the fly, learning from each iteration and improving without requiring external intervention. This adaptability makes SCoRe not only more efficient in handling computational resources but also more reliable in delivering accurate results across a wide range of applications, from coding tasks to financial modeling and beyond.

Furthermore, because SCoRe eliminates the reliance on external verifiers and reduces the need for retraining, it drastically cuts down on the time and computational power required to achieve high levels of accuracy. In fields where speed and precision are paramount, such as real-time decision-making or large-scale data processing, these efficiency gains can be transformative. AI systems powered by SCoRe can deliver accurate results faster, with fewer resources, and can do so consistently across different contexts.

In sum, SCoRe's self-correcting capabilities through reinforcement learning not only improve the overall intelligence of AI systems but also make them far more computationally efficient. By removing the need for a secondary model to verify outputs and equipping the AI with the ability to handle mismatches between training data and real-world scenarios, SCoRe reduces both computational cost and the complexity of maintaining accurate, adaptable AI systems. This innovation is key to the future of scalable AI, where the balance between performance and resource efficiency is critical for widespread adoption across industries.

Chapter 4: Real-World Results: AI Gets Smarter, Faster, and More Independent

The application of SCoRe to large language models (LLMs) like Gemini 1.0 Pro and Gemini 1.5 Flash has yielded impressive results, particularly in tasks that involve mathematical reasoning and coding. These areas traditionally challenge AI models due to their complexity and the multi-step reasoning required to arrive at correct answers. However, with the introduction of SCoRe, the performance of these models has seen a remarkable improvement, demonstrating the efficacy of self-correction via reinforcement learning.

In mathematical reasoning tasks, which often involve several stages of calculation and logical deduction, LLMs have typically struggled to maintain accuracy across multiple steps. A small error early in the process can easily cascade into a larger mistake, leading to incorrect final results. Before SCoRe, models like Gemini 1.0 Pro might have had a solid grasp of mathematical concepts but lacked the ability to detect and correct their mistakes independently. If the model miscalculated during a multi-step equation, the error would persist unless a human intervened or a second model was deployed to verify and correct the output.

With the implementation of SCoRe, however, the situation changed dramatically. The reinforcement learning process enabled the model to self-correct, leading to a 15.6% improvement in mathematical reasoning accuracy. This means that after making an initial attempt at solving a math problem, the AI could review its work, identify where it went wrong, and revise its solution in a meaningful way. In practical terms, this involved the AI making substantial corrections, rather than superficial tweaks, to ensure that the entire solution was logically consistent and mathematically sound. For example, when tackling problems from the math dataset, the model's self-correction accuracy significantly improved, allowing it to handle even complex multi-step tasks with greater precision.

This improvement in mathematical reasoning is crucial because it highlights the potential of AI systems to take on tasks that require sustained, logical thinking. In fields where accuracy in calculations is paramount—such as engineering, scientific research, or financial analysis—the ability for AI to recognize and fix its own errors offers a major advantage. SCoRe's ability to boost mathematical reasoning accuracy by such a significant margin underscores the value of reinforcement learning in elevating AI's capabilities beyond simple task execution to more complex problem-solving scenarios.

The results in coding tasks were even more dramatic. Before SCoRe, AI models often struggled with coding due to the intricate, step-by-step nature of programming. Writing code requires not only syntactic accuracy but also logical coherence. A minor error in one part of the code can render an entire program non-functional, and traditional AI models would frequently generate code that needed substantial human correction. Entering the debugging process would often involve human intervention to identify where the AI had gone wrong, whether in syntax or logic.

However, after integrating SCoRe into models like Gemini 1.5 Flash, the accuracy of coding tasks improved by an astounding 99.1%. This means that the AI could self-correct almost all of its initial mistakes when generating code, making it far more reliable for real-world applications. When applied to coding benchmarks, such as the HumanEval dataset, which tests the model's ability to write functional code, the model's performance saw a significant leap. Not only could the AI identify syntax errors, but it also became more adept at recognizing logical flaws in the code and correcting them during the self-correction phase.

This leap in coding accuracy is transformative for industries that rely on software development, where

AI is increasingly being used to generate and refine code. By enabling AI to self-correct in coding tasks, SCoRe has the potential to drastically reduce the time developers spend debugging and refining AI-generated code. This can accelerate the software development lifecycle, making it easier for teams to implement AI-generated code that is functional, syntactically correct, and logically sound. In practical terms, this improvement could mean fewer bugs, faster iteration times, and more efficient coding practices across the board.

In sum, the application of SCoRe to mathematical reasoning and coding tasks has resulted in significant gains in accuracy and reliability. The 15.6% improvement in math reasoning accuracy demonstrates AI's growing capacity to handle complex, multi-step problems, while the 99.1% improvement in coding accuracy underscores the model's ability to autonomously produce high-quality, functional code. These results are a testament to the power of reinforcement learning and self-correction in elevating AI's capabilities, particularly in tasks that require logical consistency and precision across multiple steps. As AI continues to integrate more deeply into industries that rely on mathematical and programming expertise, the advancements driven by SCoRe will only become more impactful.

SCoRe's success, while initially demonstrated through significant improvements in mathematical reasoning and coding tasks, extends far beyond these domains. Its potential to transform various fields, such as scientific research, financial modeling, and education, highlights the far-reaching implications of self-correcting AI across diverse industries. The ability of AI to independently recognize and fix its mistakes not only enhances accuracy but also opens doors for greater efficiency, scalability, and innovation across many critical sectors.

In the realm of **scientific research**, AI models equipped with SCoRe can revolutionize how data is analyzed and interpreted. Research often involves processing massive datasets and complex simulations, tasks where even minor errors can derail an entire study. Traditionally, researchers would need to manually check AI outputs or rely on secondary models to verify results. However, with SCoRe, AI can self-correct as it works through these data-intensive tasks, ensuring that mistakes are identified and rectified early in the process. This means that researchers can trust the AI to perform more independently, reducing the need for constant human oversight and increasing the speed at which results can be achieved. Whether it's modeling climate change scenarios, analyzing genomic data, or conducting particle physics simulations, self-correcting AI could accelerate breakthroughs by handling complex, multi-step tasks with fewer errors and greater autonomy.

In **financial modeling**, precision is critical. Even the slightest miscalculation can have enormous consequences, affecting everything from risk assessments to market predictions. Financial analysts rely on models to sift through large volumes of data, make predictions, and simulate future scenarios. However, these models are prone to errors, especially when the data inputs don't match the scenarios they were trained on, leading to inaccurate forecasts. SCoRe offers a solution to this by enabling AI models to detect and fix their mistakes in real time. This can drastically reduce the occurrence of costly errors, allowing financial institutions to make more informed decisions based on accurate, self-correcting models. From stock market analysis to credit risk evaluations, AI powered by SCoRe has the potential to reshape the financial landscape by providing more reliable, real-time insights with minimal human intervention.

Education is another field where SCoRe's impact could be profound. AI is increasingly being used to assist with personalized learning, providing students with tailored lessons and feedback based on their individual progress. However, one of the challenges in educational AI has been ensuring that the feedback provided is accurate and helpful, especially in complex subjects like math or science. Traditionally, educational AI systems needed to be meticulously supervised to avoid giving students incorrect or misleading information. With SCoRe, educational platforms can provide far more reliable guidance. The AI can detect when it has made an error in providing feedback or solving a problem and correct itself, ensuring that students are receiving accurate, constructive support. This not only enhances the quality of education but also allows for more adaptive, personalized learning experiences where students can engage with AI tutors that grow smarter and more effective over time.

Moreover, in **healthcare**, SCoRe could be instrumental in improving diagnostic tools and personalized treatment plans. AI models in healthcare often analyze vast amounts of patient data to suggest diagnoses or recommend treatments, but mistakes in these assessments can be life-threatening. By integrating SCoRe, healthcare AI systems can self-correct, improving their diagnostic accuracy by learning from their mistakes and refining their outputs with each patient case. This capability could lead to more reliable AI-driven healthcare systems, reducing the burden on medical professionals and ensuring better outcomes for patients.

In **automated customer service** and **natural language processing (NLP)**, SCoRe also offers great promise. AI chatbots and virtual assistants, for example, are frequently used in customer service to handle common inquiries and solve customer problems. However, these systems often fail when faced with complex or unexpected requests, requiring human intervention. SCoRe can enable such systems to learn from their mistakes, improving their ability to handle more complicated queries without needing a human to step in. Over time, this could lead to more efficient and autonomous customer service platforms, allowing businesses to streamline their operations while still providing high-quality service.

Even in **creative fields**, like art or music composition, SCoRe can help AI systems produce more innovative results. By learning from its past mistakes—whether in generating a faulty chord progression or creating an unsatisfactory visual design—the AI can refine its artistic outputs over multiple iterations. This ability to self-correct allows AI to participate more meaningfully in the creative process, working alongside human creators to bring imaginative concepts to life.

What makes SCoRe's success across these various domains so significant is its flexibility. The reinforcement learning-based self-correction process does not rely on pre-programmed rules or static data, meaning that it can be applied to a wide range of tasks that require dynamic decision-making and adaptability. Whether in industries driven by data, creativity, or human interaction, the core benefit of SCoRe is its ability to continuously improve, adapting to new challenges and scenarios without human intervention.

Ultimately, SCoRe's impact goes far beyond improving the accuracy of AI in math and programming. Its ability to extend into fields like scientific research, financial modeling, education, healthcare, and even creative industries demonstrates the transformative potential of self-correcting AI. By enabling AI to autonomously refine its outputs across different tasks and domains, SCoRe positions AI to play an even more integral role in shaping the future of these industries, driving innovation, efficiency, and accuracy at every turn.

Chapter 5: Generalizing Self-Correction Across Multiple Domains

The breakthrough achieved through Google DeepMind's SCoRe method is poised to have a profound impact across a wide array of industries, particularly those that rely on complex, multi-step reasoning. One of the most immediate and promising applications is in **software development**, where AI's ability to autonomously write, debug, and refine code can significantly enhance productivity and accuracy. Traditionally, software development is a painstaking process that requires not only writing functional code but also continuously debugging and refining it to eliminate errors. This process can be time-consuming and requires an expert understanding of both the programming language and the logic behind the software.

With SCoRe's self-correcting capabilities, AI can take on a much larger role in this process. Instead of merely generating code, AI models equipped with

SCoRe can now autonomously detect and fix errors in their own outputs, effectively debugging themselves. This goes beyond just identifying syntax errors, which many existing models can already do; it enables AI to spot logical flaws or inefficiencies in the code and revise them accordingly. For developers, this means that much of the iterative process of refining code can be handled by the AI, freeing them up to focus on more complex aspects of software design and architecture. The improvement in coding accuracy seen in the application of SCoRe—up by 99.1%—illustrates just how transformative this technology can be for software development.

By streamlining the coding and debugging process, SCoRe-enabled AI systems can dramatically shorten development cycles. Instead of developers needing to run multiple tests and manually correct errors, they can rely on AI to autonomously fix many of the issues that would otherwise slow down the project. This not only increases efficiency but also reduces the risk of human error in the debugging process. As AI becomes more adept at understanding and correcting its own mistakes, the reliability of AI-generated code will continue to improve, making it a valuable tool not just for individual developers but for large-scale software projects where speed and precision are critical.

Beyond software development, SCoRe's impact will be felt in other fields that rely on **multi-step reasoning**, where mistakes made in the early stages of a process can snowball into larger issues if not corrected in time. In industries like **manufacturing**, for example, AI systems that control production lines or optimize supply chains often face complex, layered decision-making processes. If an error occurs at any point in the chain—whether it's a miscalculation in inventory management or an incorrect adjustment to production schedules—it can cause costly disruptions. AI models using SCoRe can autonomously detect and correct these errors in real time, ensuring that processes remain efficient and accurate throughout.

Healthcare is another sector where SCoRe's ability to handle complex reasoning can have life-saving implications. AI is increasingly being used to assist in diagnostics and treatment planning, where it must process vast amounts of patient data and make multi-step decisions about the best course of action. However, traditional AI systems have struggled to correct errors without human intervention, which can lead to incorrect diagnoses or ineffective treatment suggestions. With SCoRe, healthcare AI can self-correct as it processes information, improving its recommendations and reducing the likelihood of errors. This capability is particularly valuable in areas like personalized medicine, where treatment decisions are based on a complex interplay of genetic, environmental, and clinical data. SCoRe-enabled AI can ensure that every step in the analysis process is accurate, providing more reliable and effective treatment plans for patients.

In **financial services**, AI models are increasingly used to analyze market trends, predict risks, and manage portfolios. These tasks often involve multi-step reasoning, where an incorrect assumption or misinterpretation of data can lead to significant financial losses. With SCoRe, financial AI models can dynamically adjust their predictions and strategies based on real-time feedback, ensuring that errors are corrected early in the decision-making process. For example, if an AI model misinterprets a market trend in its initial analysis, it can revise its conclusions before making any investment recommendations, reducing the risk of costly mistakes. This ability to autonomously refine its approach makes AI more reliable for tasks like risk assessment, market analysis, and even algorithmic trading, where split-second decisions can have enormous consequences.

SCoRe's potential impact extends even further into fields like **autonomous vehicles** and **robotics**, where real-time decision-making and self-correction are essential. Autonomous systems must process a constant stream of data from their environment and make split-second decisions to navigate safely and efficiently. If an error occurs—such as a misinterpretation of an obstacle or an incorrect adjustment in the vehicle's speed—the consequences can be dire. With SCoRe, these systems can recognize when a mistake has been made, correct it on the fly, and continue operating smoothly. This not only increases the safety and reliability of autonomous systems but also makes them more adaptable to unpredictable real-world conditions.

In **education**, SCoRe can enhance personalized learning platforms by enabling AI tutors to self-correct their responses based on student performance. When an AI system detects that it has given incorrect or confusing feedback to a student, it can revise its explanation, helping students grasp concepts more effectively. Over time, these AI tutors can improve their teaching strategies, becoming more adept at identifying the unique needs of each student and providing tailored, accurate guidance. This self-correcting capability can make educational AI tools more valuable in classrooms and online learning environments, where personalized attention is often hard to achieve at scale.

Even in fields like **scientific research**, where complex simulations and data analysis are key, SCoRe can play a transformative role. Scientific models often require multiple iterations to get the right results, and even small errors in early calculations can lead to incorrect conclusions. AI systems powered by SCoRe can autonomously correct these errors during the simulation process, leading to more accurate results and faster scientific discoveries. Whether it's modeling climate change, conducting pharmaceutical research, or analyzing astronomical data, SCoRe's self-correcting abilities can significantly enhance the reliability and efficiency of AI in scientific research.

In essence, SCoRe represents a leap forward in AI's capacity to tackle complex, multi-step tasks across a wide range of industries. By enabling AI to autonomously identify and fix its own mistakes, SCoRe not only improves the accuracy of AI systems but also makes them more efficient and reliable in real-world applications. As AI continues to integrate more deeply into industries that demand high levels of reasoning and precision, the advancements made possible by SCoRe will redefine the role of AI, allowing it to operate with greater autonomy, adaptability, and intelligence.

While Google DeepMind's SCoRe has already demonstrated impressive capabilities in self-correction through its two-stage training process, the potential for scaling this method beyond just two rounds of correction opens up exciting possibilities for tackling even more complex problems. The current system is designed to engage the AI in two passes: the first focusing on identifying and making meaningful corrections, and the second refining those corrections through reinforcement learning, improving overall accuracy with each iteration. However, extending this approach to multiple rounds of self-correction could significantly enhance AI's ability to handle more intricate, layered tasks.

In its current form, SCoRe allows AI to make substantial improvements in a single iteration, which is effective for many applications, particularly those involving mathematical reasoning or coding. However, as tasks become increasingly complex—such as solving multi-step scientific problems, optimizing intricate logistical systems, or handling large-scale simulations—AI may benefit from additional rounds of self-correction. In these cases, the complexity of the task might require more than two iterations for the AI to reach optimal performance. By enabling the

AI to engage in several rounds of self-correction, DeepMind could help the AI not only refine its immediate responses but also reassess earlier corrections and further improve upon them.

For example, in a highly technical environment like climate modeling or pharmaceutical research, where simulations are built on thousands of interconnected variables, extending SCoRe to multiple rounds of correction would allow the AI to iteratively refine its calculations across a broader scope. If an error occurs early in the simulation, subsequent rounds of correction could help the AI systematically revisit and adjust each step, minimizing the impact of any missteps on the final outcome. Over several rounds, the AI could approach a higher degree of precision, ensuring that each iteration brings it closer to a fully optimized solution.

This potential for multiple rounds of self-correction becomes even more relevant in tasks that require **multi-layered decision-making**. Imagine an AI managing an autonomous supply chain, where each decision—whether related to inventory management, transportation, or demand forecasting—affects all subsequent actions. One error in the initial decision-making process could ripple throughout the system, disrupting efficiency and accuracy. Extending SCoRe to more than two rounds of correction would allow the AI to systematically revise its earlier decisions, improving both short-term and long-term outcomes. With each pass, the AI would reassess its previous corrections and make even more refined adjustments, ensuring that the entire process remains optimized over time.

In addition to scaling SCoRe across more correction rounds, Google DeepMind is exploring the possibility of **unifying the two stages of training** into a more streamlined process. Currently, the two stages are distinct: the first focusing on helping the AI identify meaningful corrections, and the second utilizing reinforcement learning to refine those corrections. While this approach has proven effective, unifying these stages could make the self-correction process even more efficient. By blending the identification and refinement processes into a continuous feedback loop, the AI could simultaneously learn to correct and optimize its decisions without needing to differentiate between initial corrections and subsequent refinements.

This **unified training process** would eliminate the separation between recognizing errors and adjusting them. Instead, the AI would engage in a fluid, ongoing correction cycle, learning from each pass and immediately applying that knowledge to further iterations. This streamlined approach could reduce the time required for training and improve the AI's ability to adapt to more complex environments. For example, in a real-time application like autonomous driving, where split-second decisions can have significant consequences, a unified training process would allow the AI to continuously update its decisions, correcting errors as they occur without waiting for a secondary refinement phase.

Moreover, by unifying the two stages of training, the AI could become more responsive in dynamic environments where conditions change rapidly. In fields like financial modeling or emergency response planning, where real-world variables shift unpredictably, AI systems would benefit from being able to immediately correct their mistakes and refine their approach as new information becomes available. With a unified correction process, the AI could handle these evolving scenarios with greater agility, improving both the speed and accuracy of its responses.

Additionally, extending the correction process to multiple rounds would allow AI to **handle increasingly abstract or creative tasks**. In creative fields like music composition, visual design, or storytelling, where there's no clear "right" or "wrong" answer, AI systems would benefit from multiple iterations of self-correction to achieve more refined, nuanced results. For instance, an AI generating a piece of music might make initial corrections based on technical errors (like incorrect chords or off-tempo notes), but additional rounds of refinement could focus on improving the emotional or artistic quality of the piece, making the output feel more human-like and expressive. The same could apply to AI-generated art or narratives, where multiple correction rounds would allow the AI to continuously refine its work until it aligns with a more subjective, higher standard of creativity or coherence.

By extending the self-correction process and unifying the stages of training, Google DeepMind can push the boundaries of what AI systems are capable of achieving. This approach not only enhances the performance of AI in traditional problem-solving contexts but also opens up new possibilities in creative and abstract tasks. As AI becomes more proficient at self-correction, we may see it applied in areas that require both technical precision and creative depth, reshaping industries and expanding the role of AI in human endeavors.

The scaling of SCoRe to multiple correction rounds and the unification of its training stages represent exciting advancements in the field of AI. By refining and streamlining the self-correction process, AI will be able to tackle more complex challenges, adapt to dynamic environments, and even contribute meaningfully to creative processes—all while operating with greater autonomy and efficiency.

Chapter 6: Overcoming the Challenges

As with any advanced system, the implementation of SCoRe in AI introduces certain challenges, one of the most significant being the risk of **overcorrection**. Overcorrection occurs when an AI, in its attempt to fix a perceived mistake, alters a correct response into an incorrect one. This issue arises from the AI's lack of understanding of context or the broader scope of the problem it's trying to solve. While self-correction is a powerful tool for improving performance, there is always the potential for the AI to misinterpret a perfectly valid output as flawed, especially if the AI over-focuses on small inconsistencies that don't actually affect the correctness of the result.

For example, when handling complex tasks such as solving a math problem or generating code, an AI using self-correction might decide that a correct solution needs refinement. Instead of leaving the correct parts untouched, the AI may alter key aspects of the response, ultimately degrading the overall accuracy. In the realm of coding, this could mean the AI revises a syntactically and logically correct piece of code into one that no longer functions as intended. Similarly, in mathematical reasoning, the AI might take a valid equation and modify it unnecessarily, leading to an incorrect final answer.

Google DeepMind has recognized this risk and implemented safeguards within the SCoRe framework to minimize the possibility of overcorrection. One such safeguard is the careful **shaping of the reward system** during the reinforcement learning phase. The AI is designed to receive rewards only for corrections that genuinely improve the overall accuracy of its response. This discourages the AI from making changes that are superficial or unnecessary. Rather than simply rewarding the AI for making any correction, the system incentivizes meaningful changes—those that enhance the outcome without disrupting correct portions of the response. In other words, the AI is rewarded for improving accuracy, not just for changing something for the sake of change.

Additionally, SCoRe integrates a feedback loop that allows the AI to reassess its own edits. After making an adjustment, the AI evaluates the overall impact of the change. If the modification doesn't lead to a significant improvement, or if it makes the response worse, the AI learns not to make that particular correction in the future. This iterative learning process helps the AI distinguish between productive and counterproductive changes, reducing the likelihood of overcorrection in subsequent rounds of self-improvement.

Another concept that plays a crucial role in preventing overcorrection is **edit distance ratios**. This metric is used to measure the difference between the AI's original output and its revised response. The edit distance ratio captures how much the AI changes from one iteration to the next. In traditional models, AI might tend to stick close to its original output, making only minor, superficial edits that don't address the core issues in the response. Conversely, if the AI makes too many drastic changes, it risks overcorrecting and disrupting correct answers.

SCoRe manages this balance by encouraging the AI to make **substantial changes only when necessary**. The edit distance ratio serves as a way to track how much the AI is modifying its output, ensuring that the changes are neither too conservative nor too aggressive. By guiding the AI toward making meaningful edits that address the underlying problems, the model avoids the pitfall of making unnecessary revisions to correct responses.

For example, in a coding task, if the AI identifies a logical error early in a sequence, the appropriate response might involve making a substantial edit to correct the logic, even if the rest of the code remains functional. The edit distance ratio allows the AI to gauge how much of its original response needs to change without overcorrecting valid portions of the code. Similarly, in mathematical reasoning, if a calculation error is detected in the early steps of a multi-step problem, the AI may need to adjust only the specific calculation, leaving the broader framework of the solution intact.

By tracking the edit distance and rewarding only meaningful corrections, SCoRe ensures that the AI doesn't gravitate toward minor, inconsequential tweaks or unnecessary overhauls of otherwise correct solutions. This balance is critical to maintaining both efficiency and accuracy in the self-correction process. The AI learns to recognize when a substantial change is needed versus when smaller adjustments will suffice, thus preventing the degradation of correct answers.

The **meaningful edit framework** within SCoRe further ensures that the AI is not just making changes for the sake of improvement, but rather is developing the capacity to target and resolve core issues without disrupting correct elements of its response. This framework, combined with the feedback loop and reward shaping, provides a robust structure that guards against the risk of overcorrection while still allowing the AI to evolve and improve through self-correction.

In summary, while overcorrection remains a potential risk in any self-correcting AI system, Google DeepMind's SCoRe addresses this challenge through a combination of reward shaping, feedback loops, and edit distance ratios. These safeguards help the AI make substantial, meaningful edits only when necessary, avoiding both superficial tweaks and unnecessary disruptions to correct responses. By ensuring that the AI focuses on the right corrections, SCoRe not only improves accuracy but also maintains the integrity of the AI's decision-making process, making it more reliable across a wide range of applications.

Reward shaping is a crucial element in Google DeepMind's SCoRe method, playing a central role in guiding the AI to make corrections that are both significant and impactful. At its core, reward shaping involves carefully structuring the feedback that the AI receives during the reinforcement learning process to encourage desired behaviors. In the context of SCoRe, this means nudging the AI to focus on meaningful, core-level corrections instead of minor, surface-level changes that don't address the root causes of errors.

In traditional reinforcement learning, an AI model learns by interacting with its environment and receiving rewards based on its actions. Over time, the model adjusts its behavior to maximize these rewards. However, without proper reward shaping, an AI can easily become fixated on making small, incremental adjustments that don't significantly improve its overall performance. In the case of SCoRe, the goal is to prevent the AI from getting stuck in this cycle of superficial corrections and instead push it toward making deeper, more impactful changes when necessary.

To understand how reward shaping works, it's helpful to consider the concept of **corrective depth**. In many cases, the AI might initially spot a small inconsistency or error in its output and make a minor change to address it. However, this surface-level correction may not resolve the underlying problem, especially in complex tasks that involve multiple layers of reasoning, like mathematical problem-solving or coding. If the AI continues to make only shallow edits, it risks creating a situation where it repeatedly corrects small details without ever addressing the core issue that caused the mistake in the first place.

Reward shaping helps prevent this by ensuring that the AI is only rewarded when its corrections lead to genuine improvements in accuracy and overall outcome. Instead of providing feedback for every minor tweak, the system is designed to offer rewards when the AI makes substantial edits that actually enhance the quality of the solution. This means that the AI learns to focus on corrections that truly matter—those that change the outcome for the better—rather than making adjustments that don't move the needle in terms of performance.

For instance, in a coding task, the AI might initially identify a minor syntax error and fix it. While this correction is important, reward shaping ensures that the AI doesn't stop there if deeper logical issues still exist in the code. The AI is encouraged to look beyond simple syntax fixes and assess whether more fundamental changes are needed to ensure the entire program runs correctly. In this scenario, the AI might rewrite a larger portion of the code to resolve logical inconsistencies rather than just patching over minor errors. Reward shaping steers the AI toward these larger, more meaningful corrections by providing rewards only when the overall functionality of the program improves.

In the context of mathematical reasoning, reward shaping works similarly. The AI might solve a multi-step equation where it initially makes a calculation error in one of the early steps. If the AI focuses on only adjusting the final answer, it may correct a superficial aspect of the solution but leave the underlying math incorrect. Reward shaping encourages the AI to revisit the entire process, correcting the early misstep and ensuring that the final solution is built on a solid foundation. By providing feedback for these core-level corrections, the system teaches the AI that meaningful improvements require addressing the root cause of the problem, not just the symptoms.

Another key aspect of reward shaping is that it discourages **overcorrection**, a common risk in self-correcting AI models. Without properly structured rewards, the AI might be tempted to overcompensate for perceived errors, changing correct responses into incorrect ones in an effort to improve its output. Reward shaping helps mitigate this by emphasizing balance. The AI learns that it doesn't need to make drastic changes unless the situation warrants it. Instead, the system rewards changes that are proportional to the problem being addressed, ensuring that the AI only makes substantial corrections when necessary.

To achieve this balance, reward shaping relies on feedback loops that allow the AI to evaluate the effectiveness of its changes. After making a correction, the AI receives feedback based on whether the change improved the overall accuracy of its response. If the correction was meaningful and addressed a core issue, the AI is rewarded, reinforcing the behavior. If the change was superficial or unnecessary, the AI either receives no reward or is penalized, teaching it to avoid such corrections in the future. Over time, the AI learns to prioritize core-level improvements and avoid wasting effort on changes that don't meaningfully impact the result.

The process of reward shaping also aligns with the **edit distance ratio** mentioned earlier. This metric helps guide the AI toward making the right amount of change—neither too much nor too little. By linking rewards to the quality and depth of the corrections, the system ensures that the AI doesn't gravitate toward superficial tweaks, nor does it make unnecessary overhauls. Instead, it learns to focus on corrections that strike the right balance, fixing the underlying problem without disrupting correct parts of the response.

In practical terms, reward shaping ensures that SCoRe-trained AI models become more efficient at self-correction over time. The AI doesn't just learn to spot mistakes—it learns to assess the importance of those mistakes and make corrections that improve the overall quality of its output. This ability to prioritize meaningful corrections allows the AI to handle increasingly complex tasks with greater accuracy, efficiency, and reliability.

In summary, reward shaping is a crucial mechanism in the SCoRe method, guiding AI toward making core-level corrections that genuinely improve performance. By structuring rewards to emphasize meaningful changes, Google DeepMind has ensured that the AI focuses on addressing the root causes of errors, rather than making minor, inconsequential adjustments. This makes

SCoRe-powered AI systems not only more accurate but also more capable of solving complex, multi-step tasks where deep, impactful corrections are essential for success.

Chapter 7: The Broader Implications of Self-Correcting AI

AI's newfound ability to correct itself has the potential to revolutionize several key industries by improving accuracy, efficiency, and autonomy. In healthcare, self-correcting AI could significantly enhance diagnostic systems and treatment planning. For instance, AI systems are already used to analyze medical images like MRIs or CT scans, helping to identify conditions such as cancer or neurological disorders. However, traditional AI models can make errors in their analysis, requiring human intervention to verify or correct the diagnosis. With the ability to self-correct, AI could automatically refine its diagnoses by learning from past mistakes. If it misidentifies an anomaly in a scan, the system could adjust its process, reducing the chances of repeating the same mistake in future analyses. Over time, this could lead to more accurate diagnoses and better patient outcomes, while also reducing the burden on healthcare professionals who no longer need to oversee every step of the AI's work.

In treatment planning, self-correcting AI could continuously improve the recommendations it makes for patient care. For chronic conditions like diabetes or heart disease, where ongoing adjustments to treatment are necessary, AI could analyze a patient's real-time data and refine its suggestions for medication or lifestyle changes. If a particular treatment plan isn't yielding the desired results, the AI could adjust the plan based on updated data without waiting for a healthcare provider to intervene. This would not only personalize care more effectively but also allow for more proactive management of patient health.

The finance sector could also experience a transformation with self-correcting AI. Financial models rely heavily on accurate predictions and data analysis, but even small mistakes can lead to significant financial losses. Traditional AI systems can help analyze market trends, forecast risks, or detect fraudulent activities, but they often struggle with unexpected shifts in market behavior or novel types of fraud. With the ability to self-correct, AI models in finance could adapt to these challenges in real-time. For example, if an AI system detects an anomaly in market behavior that contradicts its predictions, it could revise its model to better account for new variables. In fraud detection, AI could learn from both false positives and undetected fraudulent transactions, continuously refining its algorithms to better distinguish between legitimate and suspicious activities. This adaptability would make financial systems more resilient and better equipped to handle the volatility of global markets.

In scientific research, where AI is used to analyze large datasets and run complex simulations, self-correction could significantly accelerate the discovery process. Research in fields like genomics, climate modeling, and pharmaceuticals often involves multi-step calculations where errors early in the process can skew results. Traditional AI systems require human researchers to monitor outputs and intervene when something goes wrong, but with self-correcting AI, the model could independently identify when it has made an error and correct it before it affects the entire experiment. For example, in climate modeling, where countless variables interact over long periods, AI could refine its predictions over time, leading to more accurate models of future climate scenarios. In genomics, AI could improve its ability to identify genetic mutations associated with diseases by learning from cases where it initially made incorrect associations. By reducing the time spent correcting errors manually, researchers could focus more on innovation and less on troubleshooting, driving faster scientific breakthroughs.

Education is another sector where AI's ability to correct itself could bring transformative changes. AI-powered educational platforms are increasingly used to personalize learning, offering students tailored feedback based on their progress. However, these systems have traditionally struggled with providing accurate feedback on complex tasks like essay writing or problem-solving in subjects like math and science. Self-correcting AI could greatly improve the quality of these platforms. For instance, if an AI tutor incorrectly evaluates a student's math solution, it could quickly correct its mistake and offer more accurate feedback in real-time. Similarly, in language learning or essay grading, the AI could adjust its feedback based on patterns it learns from previous errors, ensuring that students receive more precise and helpful guidance. This would lead to a more personalized and adaptive learning experience, helping students master difficult concepts more effectively while reducing the need for constant teacher oversight.

Overall, AI's ability to self-correct will enable systems to become smarter, more adaptive, and more reliable across industries. In healthcare, it could lead to better diagnostic accuracy and personalized treatment. In finance, it could strengthen risk management and fraud detection. In scientific research, it could accelerate discoveries by automating error correction. And in education, it could enhance the learning process by providing more accurate and timely feedback to students. This new capability marks a significant step toward AI systems that can operate with minimal human intervention, continuously learning and improving as they interact with the world around them.

The leap forward with AI's ability to self-correct marks a transformative shift in how future AI models and tools will be developed. As AI systems become more autonomous in recognizing and fixing their mistakes, they move closer to operating independently from human oversight. This shift could reshape the future of AI in profound ways, impacting the technology's role across industries and in daily life.

Traditionally, AI models have relied on a combination of vast training data, human intervention, and supervised learning to ensure accuracy and efficiency. However, with self-correcting mechanisms like SCoRe, the dependence on human input could decrease dramatically. AI's growing ability to independently identify and amend errors paves the way for models that can learn and improve continuously in real-time, without requiring constant updates or human fine-tuning. This capability opens up a new frontier for AI development—one where systems can function more like living entities, evolving based on their experiences and adapting to changing environments or data sets.

In terms of **real-world applications**, the increasing reliability of self-correcting AI will make these systems more trustworthy in areas where precision is critical, such as healthcare, finance, and autonomous systems like vehicles. For instance, in healthcare, AI systems that can autonomously improve diagnostic accuracy by learning from each interaction could reduce human error and improve patient outcomes. Similarly, in finance, the ability of AI models to refine risk assessments or market predictions without external intervention will allow institutions to navigate complex economic environments with greater confidence. The potential to reduce human oversight also means that these systems could handle larger, more complex tasks without sacrificing accuracy, making them even more indispensable.

One of the key areas where self-correcting AI could make a significant impact is in the development of **general AI**—systems that possess the ability to perform a wide variety of tasks across different domains, much like human intelligence. Currently, most AI models are highly specialized, excelling at narrow tasks like image recognition or language processing but struggling when asked to generalize their skills. With the integration of self-correction mechanisms, future AI models could learn to adapt to new domains and tasks on their own. For example, an AI system trained in natural language processing could eventually learn to handle data analysis or even creative tasks like art or music composition, by continuously refining its own understanding and capabilities across different areas. This development could mark a significant step toward the creation of AI systems that can truly think and learn like humans.

The shift toward more autonomous, self-correcting AI also raises interesting possibilities for **collaborative human-AI interactions**. As AI becomes more adept at learning from its mistakes, it could take on a greater role as a collaborative partner in creative, scientific, and technical fields. Instead of serving as mere tools, these systems could work alongside human professionals, offering suggestions, making improvements, and even generating new ideas. For example, in scientific research, AI could identify areas where experimental data might be flawed, suggest alternative hypotheses, or adjust experimental models based on real-time results, allowing researchers to focus on the higher-level aspects of discovery and innovation.

Moreover, self-correcting AI could help bridge the gap between **human cognitive limitations and AI's computational power.** One of the challenges in human-AI interaction has been ensuring that AI outputs are interpretable and reliable enough for humans to trust. As self-correction becomes more refined, AI models will be able to explain not just their outputs but also the reasoning behind their corrections, providing more transparent insights into how they arrived at a particular solution. This increased clarity could improve trust in AI systems, making them more useful in decision-making processes in sectors like law, education, and governance.

However, this leap toward independent AI also brings challenges and questions. As AI systems become more self-reliant, there will be a growing need for **robust ethical frameworks** to govern their behavior and ensure that they operate within boundaries set by humans. Developers will need to focus on creating guardrails that prevent AI from overstepping or making decisions that could have unintended consequences, especially in high-stakes environments like healthcare or autonomous vehicles. Ensuring that AI remains aligned with human values and goals will become even more important as these systems gain more autonomy.

Another exciting frontier for AI development is the potential for **cross-domain learning**. Currently, most AI models are developed for specific applications, such as diagnosing medical conditions, driving autonomous vehicles, or analyzing financial data. With self-correction, future AI models could become better at applying knowledge learned in one domain to another, creating a more unified intelligence system. For example, an AI system that has learned to correct mistakes in a medical diagnosis could transfer that learning to improve its decision-making processes in financial risk assessment. This cross-pollination of learning could create more versatile AI systems capable of tackling a wider range of problems than today's highly specialized models.

Finally, the evolution of self-correcting AI could lead to the development of systems that require **minimal human maintenance**. Today, AI models often require retraining, fine-tuning, or updating to stay relevant as data and circumstances evolve. With the ability to self-correct and adapt, future AI systems may not need this level of ongoing maintenance. They would continuously improve on their own, responding to new data and shifting requirements without human intervention. This would free up human experts from the burden of managing and maintaining AI systems, allowing them to focus on innovation and strategic decision-making.

In conclusion, the ability of AI to correct itself represents a significant advancement that will reshape how future AI models and tools are developed. As AI systems become more reliable and independent, they will play an increasingly important role in real-world applications, operating with minimal human oversight while continuously improving their performance. This leap forward promises to revolutionize industries, accelerate innovation, and bring us closer to the development of general AI capable of tackling a wide range of tasks. While challenges remain, particularly in ensuring ethical behavior and transparency, the road ahead for AI development is one filled with transformative potential.

Chapter 8: A World with Autonomous AI: What's Next?

The possibility of AI becoming an integral part of everyday life, not just as assistants but as fully autonomous, self-improving entities, is rapidly moving from science fiction to reality. As AI systems evolve with the ability to learn, adapt, and solve problems without external intervention, their role in daily life will fundamentally shift. Instead of simply following commands or executing predefined tasks, these AI systems will continuously improve their performance, learning from mistakes and adapting to new situations much like humans do.

One of the most immediate and visible impacts of this shift will be seen in **smart home technologies**. Currently, AI-powered devices like smart speakers, thermostats, and security systems function primarily as assistants, responding to voice commands or preset instructions. While convenient, these systems are limited by their inability to adapt beyond their programmed responses. However, with the capability to self-correct and learn, the next generation of smart home AI could autonomously optimize the environment based on an individual's habits, preferences, and real-time data. For example, a smart home system might learn to adjust the lighting, temperature, or security settings based on how you interact with your home over time. It could detect patterns—such as when you're likely to return home, when to dim the lights for relaxation, or even when to activate security systems based on subtle cues—and adapt without needing specific commands.

In the realm of **transportation**, self-correcting AI could revolutionize how we interact with autonomous vehicles. Self-driving cars, which are already making strides in the transportation industry, could benefit immensely from AI systems that continuously learn and improve their driving capabilities. A car equipped with self-correcting AI could not only navigate more effectively but also adapt to new road conditions, learn from past driving experiences, and handle unforeseen situations without human intervention. For example, if the AI encounters a complex traffic scenario or an unexpected road hazard, it could refine its response for future encounters, continuously improving its driving performance. This ability to learn from real-world experiences would make autonomous vehicles far safer, more reliable, and more efficient, potentially reducing accidents and making transportation more accessible.

Beyond transportation, autonomous AI could play a significant role in **health and wellness management**. Wearable health devices like fitness trackers and smartwatches are already helping people monitor their health by tracking metrics such as heart rate, physical activity, and sleep patterns. Currently, these devices rely on pre-set algorithms to provide feedback, but they don't adapt significantly over time. With self-improving AI, however, these systems could become much more personalized and proactive. A fitness tracker, for instance, could learn from an individual's unique physiological responses, making more accurate suggestions for exercise or recovery based on long-term data. If the AI detects patterns indicating potential health risks, such as irregular heart rhythms or abnormal blood pressure fluctuations, it could refine its monitoring and provide tailored recommendations. In the future, this kind of AI could even communicate with healthcare providers, ensuring that users receive the most relevant advice based on their personal health data, all without needing external reprogramming or intervention.

AI's autonomous intelligence could also significantly enhance the experience of **personalized learning**. Educational AI systems, which are currently used to provide basic feedback and individualized instruction, could become far more dynamic and responsive to a student's needs. Instead of following a fixed curriculum or offering standardized assessments, a self-correcting AI tutor could learn how each student best absorbs information and adjust its teaching methods accordingly. If a student struggles with a particular concept, the AI could detect the issue, adapt its approach, and offer alternative explanations or resources that suit the student's learning style. Over time, this AI would become more in tune with the student's progress and gaps in knowledge, creating a truly personalized learning experience that evolves as the student grows.

In **consumer services**, AI that continuously learns and adapts could transform the way we interact with businesses. Imagine an AI-driven customer service system that not only answers inquiries but improves its problem-solving abilities with each interaction. Over time, this system could learn to anticipate customer needs, offering solutions before problems even arise. It could understand the preferences of individual customers, providing a more personalized and efficient experience that reduces frustration. In industries such as banking or e-commerce, this would create a seamless and intuitive customer experience, where AI handles more complex tasks autonomously and leaves human agents to focus on higher-level interactions.

The potential for **entertainment** is equally exciting. AI that can adapt and self-correct could revolutionize interactive experiences such as gaming or virtual reality. For example, in video games, an AI-driven system could analyze a player's style and skill level, adjusting the game's difficulty, storyline, or challenges to create a more immersive and personalized experience. The AI could learn what engages the player the most, whether it's strategic challenges, fast-paced action, or narrative-driven content, and adapt accordingly. In virtual reality environments, self-improving AI could make worlds more dynamic and responsive, offering richer, more immersive interactions based on how users engage with the environment.

Moreover, as AI becomes more autonomous, its role in **environmental sustainability** could expand significantly. AI systems could manage energy consumption in cities, homes, and businesses by learning and predicting usage patterns, optimizing energy distribution without human oversight. These AI-driven systems could reduce waste by autonomously adjusting to real-time data on weather, energy supply, and demand. In agriculture, self-correcting AI could be applied to smart farming practices, adjusting irrigation, fertilization, and pest control dynamically based on the needs of crops and the surrounding environment, improving yields and reducing environmental impact.

The integration of autonomous, self-improving AI into daily life opens up a world where machines do more than execute commands; they become capable of independent thought and decision-making, continuously learning from interactions and improving their performance. These systems will be deeply embedded into the fabric of everyday experiences, from the devices we use to the environments we live in, making our lives more efficient, personalized, and responsive. With each interaction, AI will adapt to our unique needs, preferences, and behaviors, evolving into more intelligent companions that help solve problems, enhance comfort, and anticipate challenges before they arise.

However, as AI becomes more self-reliant, it raises important questions about control, ethics, and trust. While the idea of AI operating independently is appealing in many ways, it also necessitates rigorous frameworks to ensure that these systems remain aligned with human values and safety. Developers will need to ensure that autonomous AI systems are transparent, accountable, and capable of making decisions that benefit users without overstepping boundaries. Privacy and data security will also be key considerations as AI becomes more deeply integrated into our personal lives.

Looking ahead, the innovations from Google DeepMind in the realm of self-correcting intelligence signal the dawn of a new era for AI. With the development of systems like SCoRe, which enable AI to learn from its own mistakes and continuously improve, we are only beginning to see the vast potential of these advancements. As AI becomes more autonomous, adaptive, and capable of handling increasingly complex tasks without human intervention, it is clear that Google DeepMind is positioning itself at the forefront of this evolution, pushing the boundaries of what AI can achieve.

One of the most exciting prospects for the future is the continued refinement and expansion of self-correcting AI, particularly in how it can be applied to more abstract, creative, and decision-making processes. DeepMind's focus on enabling AI to generalize across various domains could lead to the development of systems that are not just task-specific but are capable of operating across a wide range of industries and scenarios. We may soon see AI that can seamlessly transition from solving technical problems in software development to making strategic business decisions, all while learning and improving with each experience.

In terms of **general artificial intelligence** (AGI), Google DeepMind's innovations in self-correction bring us closer to creating AI that can understand, reason, and learn in a more human-like manner. By developing systems that are not limited to static data sets or pre-defined rules, but instead learn dynamically through interaction with their environment, DeepMind is paving the way for AI that can handle uncertainty, ambiguity, and the kinds of open-ended problems that currently require human intelligence. This could revolutionize fields like scientific research, where AI could generate hypotheses, design experiments, and even make discoveries with minimal human guidance. In healthcare, AGI could lead to a new generation of diagnostic tools that continuously refine their knowledge base, improving patient outcomes through real-time learning and adaptation.

Moreover, DeepMind's work is likely to play a pivotal role in advancing **multi-agent AI systems**, where multiple AI entities work together or independently within a shared environment. In such systems, AI would need to not only correct its own mistakes but also learn how to collaborate with other AI agents, balancing competition and cooperation. This kind of development could have far-reaching implications for industries like logistics, where fleets of autonomous vehicles need to coordinate complex tasks, or in finance, where trading algorithms operate simultaneously in fast-paced markets.

As AI systems become more capable of self-regulation, **ethical AI** will take center stage. DeepMind's commitment to building safe and ethical AI will be crucial as the technology becomes more widespread. Future innovations will need to address concerns around bias, transparency, and decision-making, ensuring that self-correcting AI remains aligned with human values. This might involve building in mechanisms for continuous ethical evaluation, where AI not only learns to improve technically but also becomes better at making decisions that are fair, inclusive, and aligned with societal norms.

One of the major developments on the horizon is likely to be the integration of **AI and robotics**. DeepMind's advancements in self-correcting intelligence could play a significant role in robotics, where machines need to adapt to complex and dynamic environments. Robots equipped with self-improving AI could learn from their interactions in the physical world, correcting for errors in navigation, manipulation, or perception. This would be a game-changer in industries ranging from manufacturing and logistics to eldercare and domestic robotics, where adaptability and precision are essential.

In addition to these specific applications, DeepMind's future innovations are likely to push the boundaries of **AI-human collaboration**. As AI systems become better at self-correction, they will also become more intuitive and reliable partners for human users. This opens up new possibilities for co-creativity, where AI assists with problem-solving and ideation, generating insights or novel solutions in collaboration with humans. In creative fields like design, writing, or art, self-correcting AI could be a partner in the creative process, refining its output based on feedback and producing work that feels increasingly human in quality and originality.

Looking further ahead, DeepMind's work will likely play a significant role in addressing global challenges. From climate change to healthcare access, self-correcting AI has the potential to tackle some of the most pressing problems of our time. In the realm of **sustainability**, for instance, AI systems that can autonomously optimize resource usage, reduce waste, and improve energy efficiency could be critical in helping industries and governments meet their sustainability goals. By continuously learning and improving, these systems could make real-time adjustments that balance economic growth with environmental protection, providing solutions that adapt to the needs of a changing world.

In conclusion, the future of self-correcting AI is one of boundless potential, and Google DeepMind is at the cutting edge of this evolution. By developing AI that can not only perform tasks but learn, adapt, and improve autonomously, DeepMind is setting the stage for the next generation of intelligent systems. These systems will be more reliable, versatile, and integrated into all facets of life, from healthcare and education to finance and entertainment. As AI continues to evolve, DeepMind's innovations will undoubtedly play a key role in shaping a future where machines can think, learn, and collaborate with humans in ways that were once the stuff of imagination. The path ahead is filled with exciting possibilities, and as we move forward, AI will become an even more indispensable part of how we live, work, and solve the world's most complex challenges.

Conclusion

The rise of self-correcting AI marks a monumental leap forward, one that goes beyond simply improving efficiency. This breakthrough fundamentally changes how we interact with, develop, and deploy intelligent systems, enabling machines to evolve, adapt, and grow smarter with each interaction. No longer are AI models static tools that rely on predefined rules or constant human oversight. They are becoming autonomous entities capable of learning from their own mistakes and continuously improving their performance without external guidance. This ability to self-correct opens up a new frontier, where AI can handle increasingly complex and dynamic challenges across industries, from healthcare and finance to education and scientific research.

The impact of self-correcting AI is profound. In healthcare, these systems can revolutionize diagnostics and personalized treatment plans by providing more accurate and real-time data analysis, improving patient outcomes. In finance, self-improving models can navigate unpredictable market conditions, enhancing risk management and fraud detection. For education, AI's capacity to adapt and refine its instruction could create highly personalized learning experiences, transforming how students engage with and absorb information. Scientific research will be accelerated, with AI helping researchers tackle problems faster by correcting errors in real time. This technology is not merely a tool for optimizing existing processes; it represents a complete reimagining of how AI will integrate into our daily lives, industries, and society.

As we stand on the cusp of this new era, it is essential to reflect on the far-reaching implications of self-correcting AI. This technology is not just reshaping businesses and industries; it is redefining our relationship with machines and technology. AI is becoming an active participant in decision-making, creativity, and problem-solving, blurring the line between human intelligence and artificial intelligence. With AI that can adapt and improve, we move closer to creating systems that collaborate with humans, anticipate needs, and evolve alongside us.

The societal implications of this shift are immense. As AI takes on more complex roles, we must consider the ethical, social, and regulatory frameworks needed to ensure that these systems serve the greater good. Questions about transparency, accountability, and fairness will become even more pressing as AI gains autonomy. It is essential that as we embrace this powerful technology, we also ensure that it remains aligned with human values, ethical standards, and societal goals.

In conclusion, the AI revolution is just beginning. Self-correcting AI is not merely an incremental advancement; it is a transformative step toward building systems that can think, learn, and grow autonomously. This innovation will drive profound changes across industries and societies, offering unprecedented opportunities for growth, creativity, and problem-solving. As we look to the future, we must embrace both the potential and the responsibility that comes with this new era of AI. The road ahead is filled with possibilities, and as these intelligent systems continue to evolve, they will reshape the world in ways we are only beginning to imagine.

www.ingramcontent.com/pod-product-compliance
Lightning Source LLC
Chambersburg PA
CBHW050320230526
45471CB00005B/2269